On the Moon

Experts on child reading levels
have consulted on the level of text and
concepts in this book.

At the end of the book is a "Look Back and Find" section
which provides additional information and encourages
the child to refer back to previous pages
for the answers posed.

Angela Grunsell trained as a teacher in 1969.
She has a Diploma in Reading and Related Skills
and for the last five years has advised London
teachers on materials and resources.

Published in the United States in 1983 by
Franklin Watts, 387 Park Avenue South, New York, NY 10016

© Aladdin Books Ltd/Franklin Watts

Designed and produced by
Aladdin Books Ltd, 70 Old Compton Street, London W1

ISBN 531-04631-1

Printed in Belgium

All rights reserved

FRANKLIN · WATTS · FIRST · LIBRARY

On the Moon

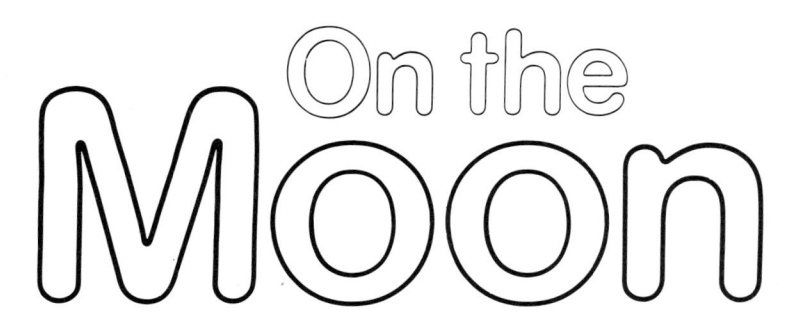

Consultant
Angela Grunsell

Illustrated by
Tessa Barwick and Elsa Godfrey

Editor
Jenny Vaughan

Franklin Watts
New York · London · Toronto · Sydney

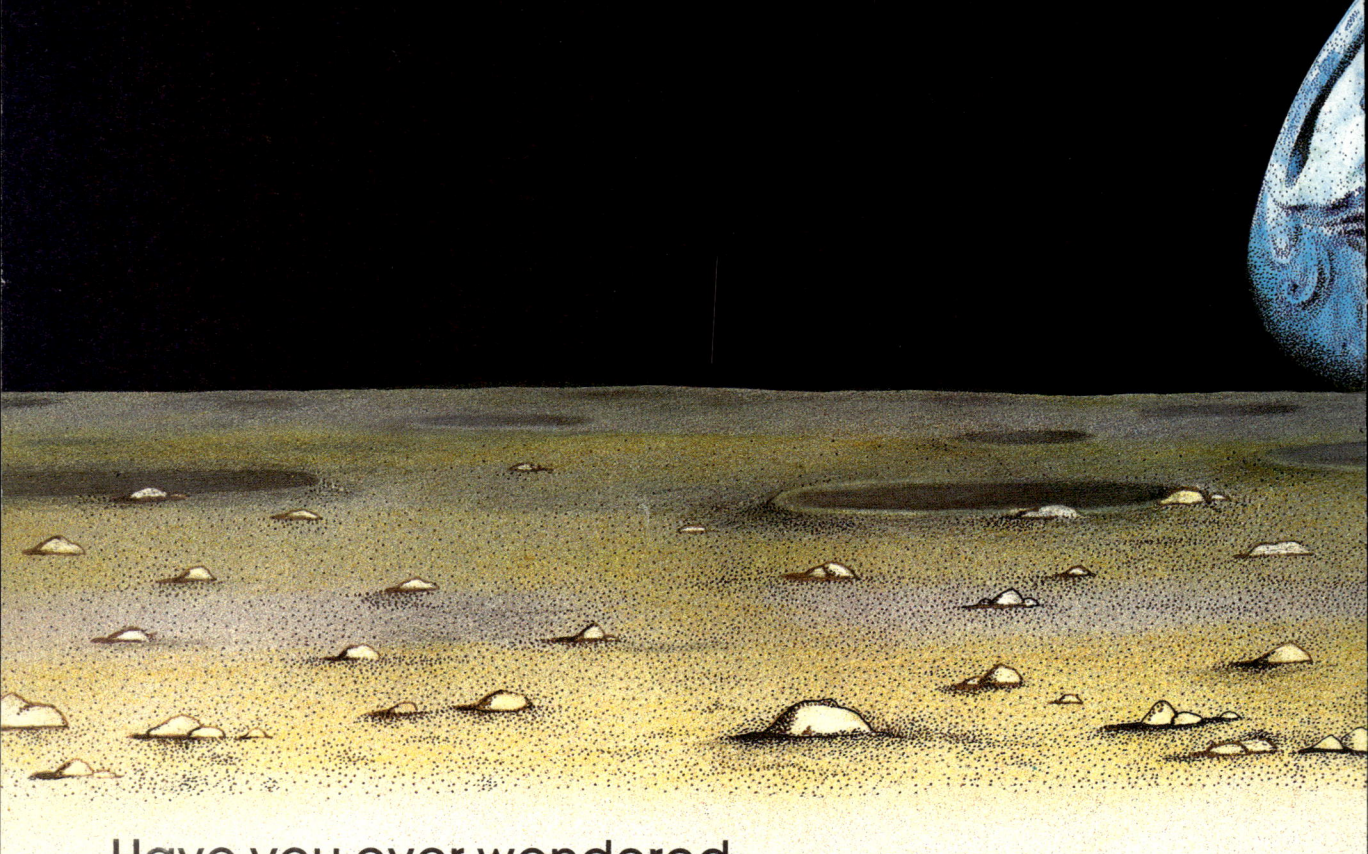

Have you ever wondered
what it's like on the Moon?
It looks like this.

When you're on the Moon
you can see the Earth in the sky.
How did people get to the Moon?

These three Americans went to the Moon in a spacecraft, in 1969.
They are called astronauts.
They were the first people to go to the Moon.
A rocket blasted the spacecraft into space.

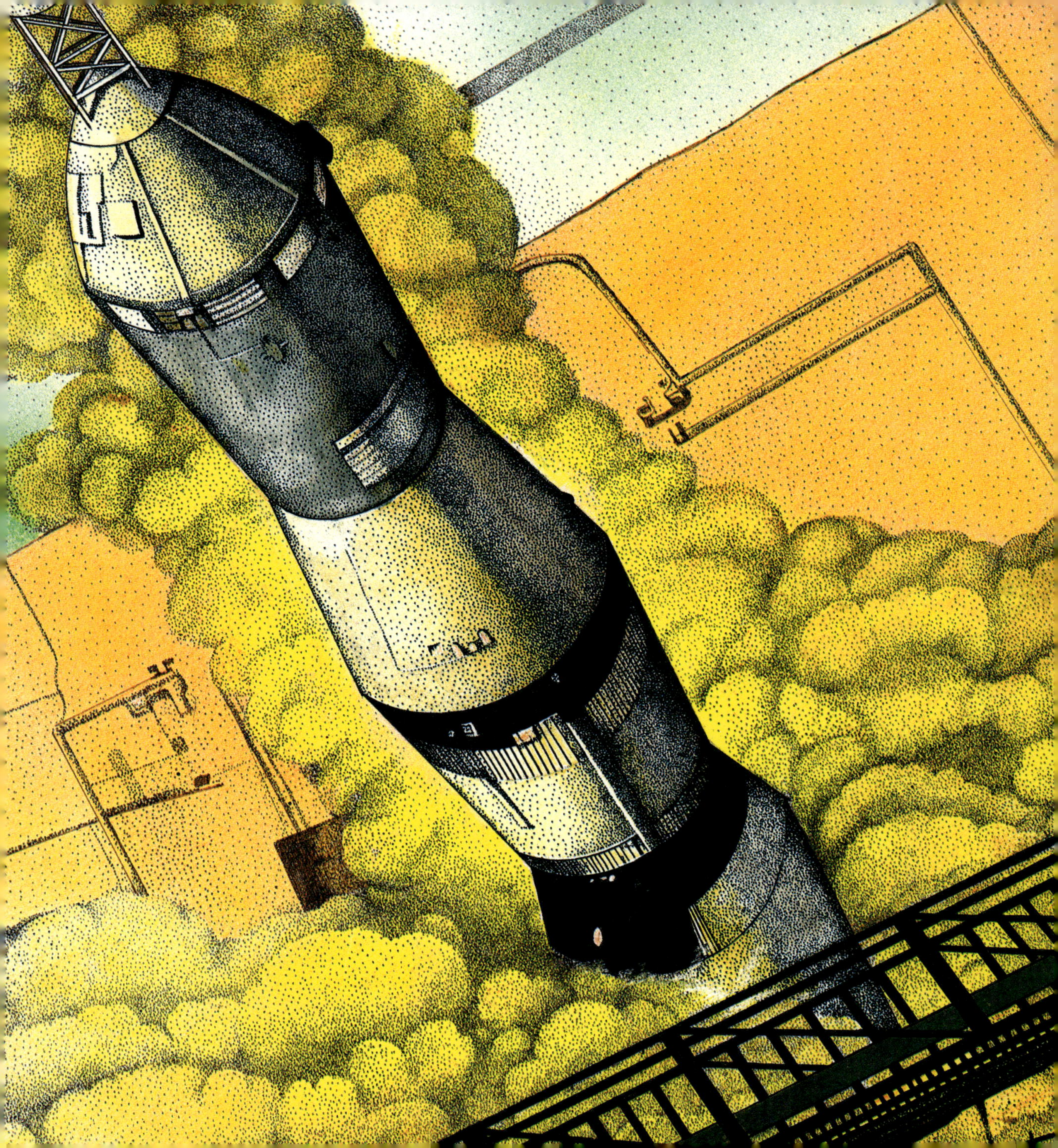

Out in space, the rocket fell away from the spacecraft in three stages. This shows the second stage.

The top of the spacecraft turned front to back.
Now the lunar module was in the right position.
The spacecraft circled the Moon in an orbit.

The astronauts seemed to have no weight.
They floated around inside the spacecraft.

They had to strap themselves to their seats to sleep.

Two of the astronauts used the lunar module
to land on the Moon.
The other one stayed behind in the spacecraft.
He continued to orbit the Moon.

The first man to step on the Moon was Neil Armstrong.

There is no wind or rain on the Moon.
The astronauts' footprints
may stay there forever.

A Moon day lasts for fourteen Earth days.
It is very hot. The night is just as long.
It is very, very cold.

The astronauts walked around on the Moon. They found they could jump much higher than they could on Earth. They also found they could lift huge rocks easily.

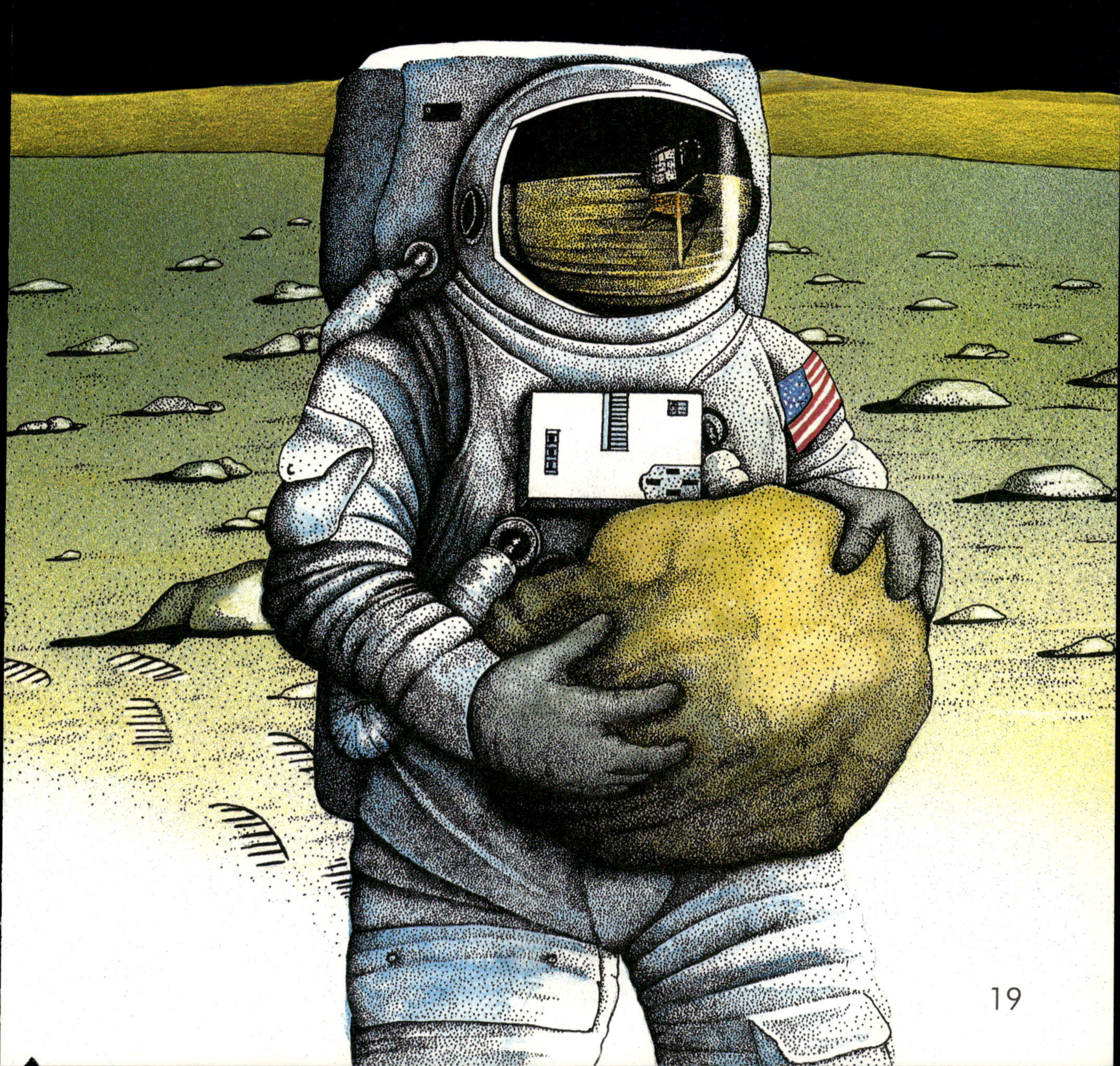

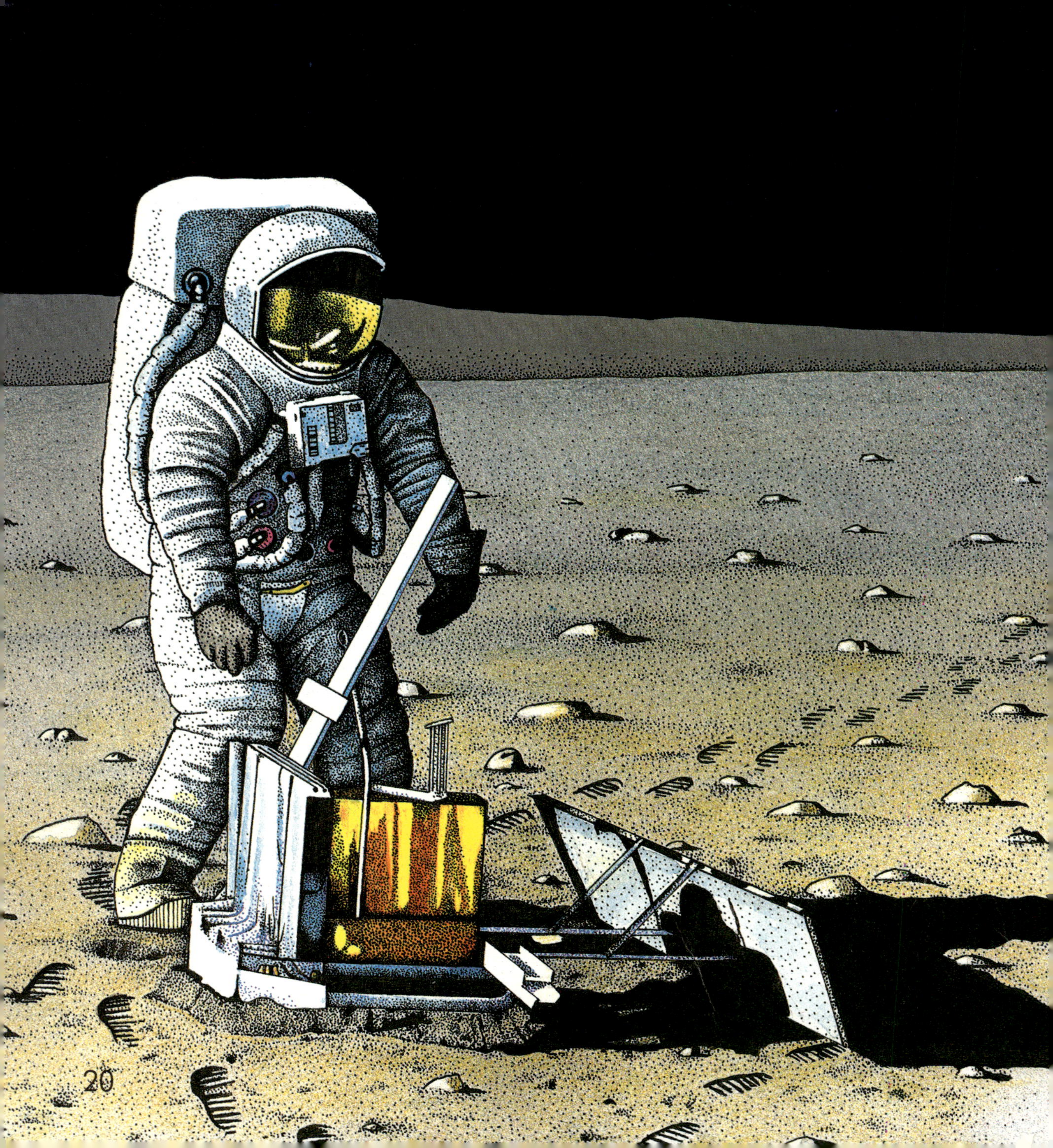

The astronauts took dust and rocks with them.
They left behind instruments that
could send information to Earth.

The lunar module blasted off from the Moon and returned to the spacecraft.
The module joined up with the spacecraft.

The three astronauts were together again and traveled back to Earth.
They landed safely in the sea.

Astronauts who went to the Moon on later trips took a special vehicle with them.
It was called a lunar rover or moon buggy.
They traveled in the buggy and saw more of the Moon.

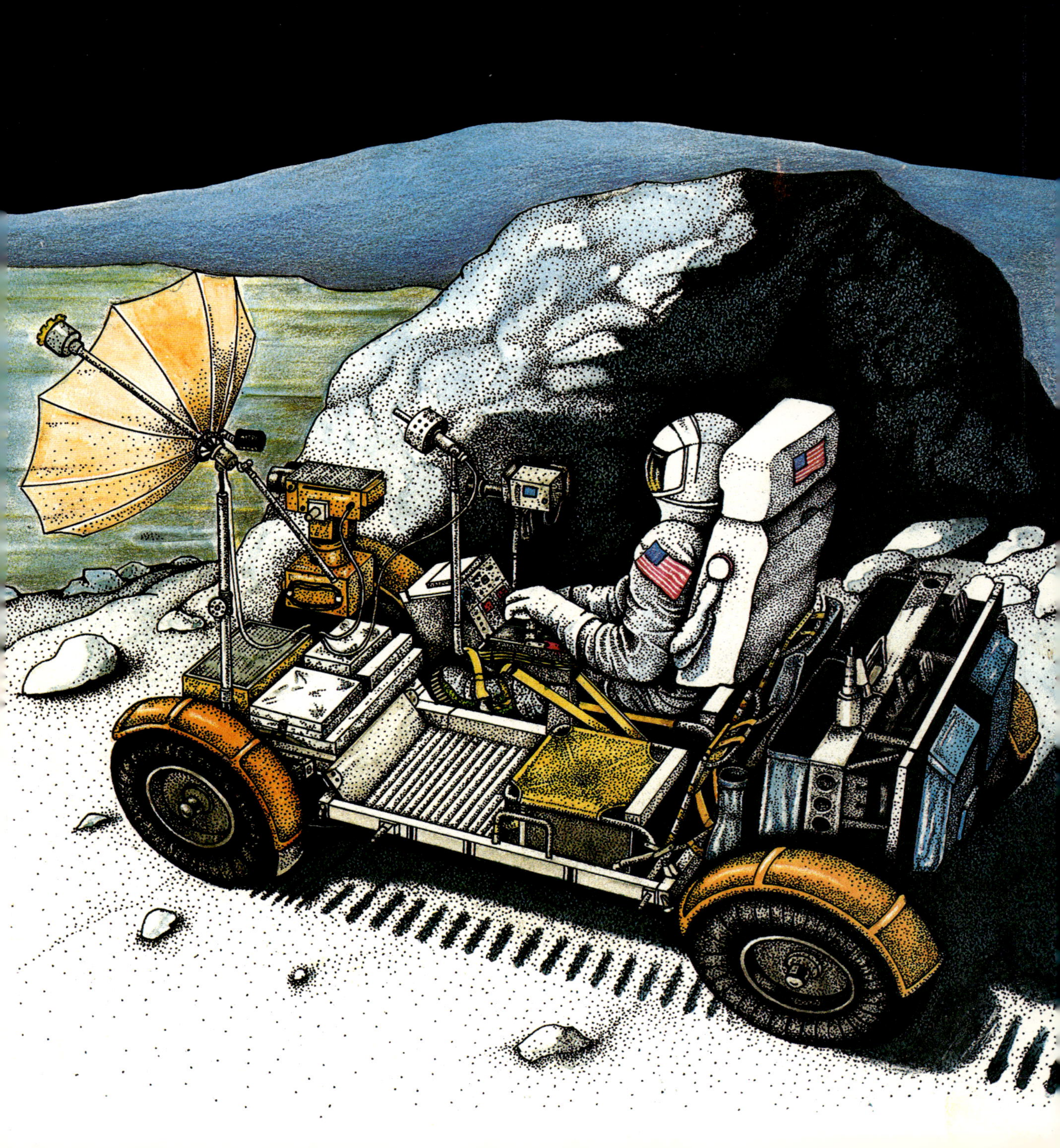

One day it may be easier to get to the Moon. Perhaps scientists will want to work there.

They might build a space station.
What do *you* think it will look like?

Look back and find

Who were the first people to go to the Moon?
They were Edwin Aldrin, Michael Collins and Neil Armstrong.

Who was the first man to walk on the Moon?
When did this happen?

What is this called?
It is part of the Saturn V rocket.

What was it used for?
What finally happened to it?

How long did it take the astronauts to get to the Moon from Earth?
It took them four days.

How did the two astronauts get from the spacecraft to the Moon?
What did the third astronaut do?

Why did they have to wear spacesuits?
*To protect themselves from
the extreme temperatures and
different atmosphere on the Moon.*

What else is different
about conditions on the Moon?

Why did the astronauts leave a flag
on the Moon?
*So that later astronauts would know which
country first landed on the Moon.*

The flag was supported by a wire frame. Why?
Which flag is it?

How was this vehicle powered,
and how fast could it go?
*It was electrically powered
by two batteries and it could go at 8½ mph*

What is it called?
Why did later astronauts take it with them?

Index

A   Aldrin, Edwin 28
Americans 8
Armstrong, Neil 15, 28
astronauts 8, 12, 14, 16, 19, 23, 24, 28, 29
atmosphere 29

B   batteries 29

C   Collins, Michael 28

D   dust 21

E   Earth 7, 17, 19, 21, 23, 28

F   flag 29

I   instruments 21

L   lunar module 11, 14, 22, 24
lunar rover 24

M   Moon 6, 7, 8, 11, 14, 15, 16, 17, 19, 22, 24, 25, 28, 29
moon buggy 24

O   orbit 11, 14

R   rain 16
rocket 8, 10, 28
rocks 19, 21

S   Saturn V rocket 28
spacecraft 8, 10, 11, 12, 14, 22, 28
space station 26
spacesuits 29

T   temperatures 29

V   vehicle 24, 29

W   weight 12
wind 16

PRINTED IN BELGIUM BY
proost
INTERNATIONAL BOOK PRODUCTION